Hacking With Kali Linux

The Complete Beginner's Guide about Kali Linux for Beginners

(Step by Step Guide to Learn Kali Linux for Hackers)

Michael Lee

Published By **Simon Dough**

.

Michael Lee

Hacking With Kali Linux: The Complete Beginner's Guide about Kali Linux for Beginners (Step by Step Guide to Learn Kali Linux for Hackers)

ISBN   978-1-998901-86-9

Legal & Disclaimer

TABLE OF CONTENTS

Introduction

Hackers work with the computer or program code, which is a set of instructions that work in the background and make up the software. While a lot of hackers do know how to program code, many downloads and use codes programmed by other people. The main requirement to know is how to work this code and adjust it to their advantage. For malicious hackers, that can be using it to steal passwords, secrets, identities, financial information, or create so much traffic that the targeted website needs to shut down.

Stealing passwords

Passwords are easy to hack because humans are very predictable. We think we are unique until it comes to passwords, but we are very easy to guess. For example, women will often use personal names for passwords—think kids, relatives, old flings—while men will stick to hobbies. The numbers we use most frequently are 1 and 2, and they are most often placed at the end of our password. More often than not, we use one word followed by some number, and if the website insists on including a capital letter, we place it at the beginning of the word and then whine about how this website is so annoying for making us go through all of this.

But how do hackers access our passwords? Well, there are several useful techniques.

The trial and error technique is called the brute force attack, and it is when you try possible combinations of letters and words to try and guess the right password. This can work because, as previously mentioned, we are very predictable when it comes to the type of passwords we use.

Another similar technique is called the dictionary attack; hackers use a file containing all the words that can be found in the dictionary, and the program tries them all. This is why it is often suggested to add numbers to your passwords as words, but this doesn't mean your "sunshine22" password is hackerproof.

A third technique is the rainbow table attack. The passwords in your computer system are hashed (generated into a value or values from a string of text using a mathematical function) in encryption. Whenever a user enters a password, it is compared to an already stored value, and if those match, you are able to enter into the website or application. Since more text can produce the same value, it doesn't matter what letters we input as long as the encryption is the same. Think of it as a door and a key. You enter the doors with the key made for that lock, but if you're skilled at lock picking or a locksmith, you don't need that exact key to enter.

How to protect yourself from password attacks

Use the salt technique. This refers to adding simple random data to the input of a hash function. The process of combining a password with a salt which we then hash is called salting. For example, a password can be "sunshine22" but adding the salt is e34f8 (combining sunshine22 with e34f8) makes your hash-stored, new salted password "sunshine22e34f8." The new salted password is thus hashed into the system and saved into the database. Adding the salt just lowered the probability that the hash value will be found in any pre-calculated table. If you are a website owner, adding salt to each user's password creates a much more complicated and costly operation for hackers. They need to generate a pre-calculated table for each salted password individually, making the process tedious and slow.

Even with the salt technique, determined hackers can pass through the "password salting." Another useful technique is the peppering technique. Just like the salt, pepper is a unique value. Pepper is different than salt because salt is unique for each user, but pepper is for everyone in the database. Pepper is not stored in the database; it's a secret value. Pepper means adding another extra value for storing passwords.

For example, let's say the pepper is the letter R. If the stored password is "sunshine22," the hash stored will be the hashed product of "sunshine22" with the added letter R. When the user logs, in the password they are giving is still "sunshine22," but the added pepper is storing "sunshine22" with the added R. The user has no knowledge that pepper is being used. The website will then cycle through every possible combination of peppers, and by taking upper and lowercase letters, there will be over 50 new combinations. The website will try hashing "sunshine22A," "sunshine22B," and so on until it reaches "sunshine22R." If one of the hashes matches the stored hash, then the user is allowed to log in. The whole point of this is that the pepper is not stored, so if the hacker wants to crack the password with a rainbow table or dictionary attack, it would take them over 50 times longer to crack a single password.

Phishing attacks

The easiest way to get someone's password is to ask them. After all, why bother with all the algorithms and cracking codes when you can just politely ask?

Phishing is often a promise of a prize if you click on a certain link that then takes you to a fake login page where you simply put in your password. The easiest way to defend from this is smart clicking, or not clicking on scammy pop-up ads.

Vacations and iPods are not just given away with a click and "you won't believe what happened next" is a sure sign of a clickbait leading to phishing.

Miracle weight loss pills, enlargement tools, singles waiting to meet you in the area and other promises of luxurious life with just one click are all phishing. Unfortunately, we have to work for money and workout for weight loss.

Chapter 1: Hacking with Kali Linux

Kali Linux is an operating system that has many tools that are supposed to be utilized by security experts. There are more than 600 tools that have been pre-installed in the operating system. In this chapter, the main discussion will be about the Kali Linux for beginners. There are many people who may not be conversant with Kali Linux; however, in this chapter, it will be possible to learn how you can easily maneuver Kali Linux.

Kali Linux

The BackTrack platform was mainly formed for the security professionals and there are many tools that had been pre-installed in the Operating System. Since Kali Linux is the predecessor of the platform, the operating system also has many tools that can be utilized by the security professionals. The tools are mainly to be used by professionals such as network administrators and the security auditors. When using these tools, it is

7

possible to assess the network and also ensure that it is secure. There are different types of hackers and they all have access to these tools.

BackTrack was useful to the security professionals, however, the main issue was that the architecture was quite complex and the tools that have been pre-installed could also not be used easily. The tools were present in the pentest directory and they were very effective when carrying out a penetration test. Many subfolders were also present and most of the tools could also be detected easily. The tools that were available in the platform include sqlninja- the tool comes in handy when carrying out SQL injection. Many more tools are also available and they can also be used to perform web exploitation when you are assessing the vulnerabilities that are present in the web applications.

Kali Linux replaced BackTrack and the architecture of the operating system is based on the Debian GNU, and it adheres to the

8

Filesystem Hierarchy System (FHS) which also has many advantages as compared to the BackTrack platform. When you use the Kali Linux operating system, you can access the available tools easily since some of the applications can be located in the system path.

Kali Linux offers the following features:

The operating system supports many desktop environments such as XFCE, Gnome, KDE, and LXDE. The operating system also offers some multilingual support.

The tools offered by the operating system are Debian-compliant and they can also be synchronized at least four times daily using the Debian repositories. The packages can also be updated easily while also ensuring that some security fixes have also been applied.

Kali Linux allows ISO customization and that means as a user, you can come up with different versions of Kali Linux that suit your needs.

The operating system has both ARMEL and ARMFH support and that means that the users can also be able to install the Kali Linux operating system in different devices.

The tools that have been pre-installed also have some diverse uses.

Kali Linux is open source and that means it is free.

In this chapter, the main focus will be on the Kali Linux operating system as a virtual machine. For starters, the main focus will be on the Kali Linux for beginners to ensure that as a reader, you can get an overview of the operating system. To use an operating system as a virtual machine, you should utilize the VMware and that means that the Kali Linux operating system will be running on the "Live Mode."

There is a reason why the VMware is used and it is because it is easy to use and it comes in handy, especially when you execute different applications that are located in the primary operating system. For example, when you install the "Live Mode" on any operating system, you can use the applications that are present in the operating system. Additionally, you can retrieve the results that you have obtained when you carry out penetration testing using the virtual machine. The test results will allow you to learn about the vulnerabilities that are present in the system.

When you launch the Kali Linux operating system, the default desktop will appear and you will also notice that there is a menu bar as well as different icons. After selecting the menu item, you will be able to gain access to numerous security tools that have been pre-installed in the operating system.

How to Configure Secure Communications

When you use Kali Linux, you must ensure that there is connectivity to a wired or wireless network. After ensuring there is connectivity, the operating system will be able to handle various updates. Also, you can customize the operating system as long as there is connectivity. First, make sure there is an IP address. After that, confirm the IP address using the ifconfig command. You can confirm it using the terminal window and an example of the command being executed is as shown below:

In this case, the IP address is 192.168.204.132. At times, you may not be able to obtain the IP address and that means that you should use the dhclient eth0 command. The DHCP protocols will issue the IP address. Other interfaces can also be used to obtain the IP address and it will depend on the configurations that are present within the system.

When using a static IP address, you can also provide some additional information. For example, you can use the following IP address in such a manner:

After opening the terminal window, make sure that you have keyed in the following command:

Make sure that you have noted the changes that have been made to the IP settings. The changes will also not be persistent and they will not reappear after you have rebooted the operating system. In some instances, you may want to make sure that such changes are permanent. To do so, ensure that you have edited the /etc/network/interfaces file. The screenshot below can offer some subtle guidance:

When you start the Kali Linux operating system, the DHCP service will not be enabled. You are supposed to enable the DHCP service automatically. After enabling the service, the new IP addresses within the network will also be announced and the administrators will also receive an alert that there is an individual carrying out some tests.

Such an issue is not major; nonetheless, it is advantageous for some of the services start automatically in the process. Key in the following commands so that you may be able to achieve all this.

When using Kali Linux, you can also install varying network services including DHCP,

HTTP, SSH, TFTP, and the VNC servers. The users can invoke these services straight from the command-line. Also, users can access these services from the Kali Linux menu.

Adjusting the Network Proxy Settings

Users can use proxies that are authenticated or unauthenticated and they can modify the proxy settings of the network using the bash.bashrc and apt.conf commands. The files will be present in this folder- /root/etc/directory.

1) Edit the bash.bashrc file first. A screenshot will be provided below since it will come in handy when offering some guidance. The text is also useful in such instances, especially if you want to add lines to the bash.bashrc file:

2) The proxy IP address will then be replaced with the Proxy IP address that you're using. Also, you will have administrator privileges and that means that you can also change the usernames and the passwords. In some instances, you may also have to

14

perform some authentication and you must key in the '@'symbol.

3)   Create an apt.conf file in the same directory while also entering the commands that are showcased in the following screenshot:

4)   Save the file and then close it. You can log in later so that you can activate the new settings.

Using Secure Shell to Secure Communications

As a security expert, you must ensure that the risk of being detected is minimized. With Kali Linux, you will not be able to use the external listening network devices. Some of the services that you can use are such as Secure Shell. First, install Secure Shell and then enable it so that you can use it.

The Kali Linux has some default SSH keys. Before starting the SSH service, disable the default keys first and also generate a keyset that is also unique since you may need it at some point. The default SSH keys will then be

15

moved to the backup folder. To generate the SSH keyset, use the following command:

To move the original keys, you should use the following command. Also, you can generate some new keysets using the same command.

Make sure that each of the keys has been verified. You can verify each key by calculating the md5sum hash values of every keyset. You can then compare the results that you have with the original keys.

When you start the SSH service, start with the menu and then select the Applications- Kali Linux- System Services-SSHD- SSHD start.

It is also possible to start the SSH when you are using the command line and this screenshot will guide you:

To verify that the SSH is running, execute the netstat query. The following screenshot will also guide you:

To stop the SSH, use this command:

Updating Kali Linux

For starters, the users must patch the Kali Linux operating system. The operating system must also be updated regularly so that it may also be up to date.

Looking into the Debian Package Management System

The package management system relies on the packages. The users can install and also remove packages as they wish when they are customizing the operating system. The packages support different tasks such as penetration testing. Users can also extend the functionality of the Kali Linux such that the operating system can support communications and documentation. As for the documentation process, run the wine application so that you can run applications such as the Microsoft Office. Some of the packages will also be stored in the repositories.

Packages and Repositories

With Kali Linux, you can only use the repositories provided by the operating system. If the installation process has not been completed, you may not be able to add the repositories. Different tools are also present on the operating system, although they may not be present in the official tool repositories. The tools may be updated manually and you should overwrite the packaged files that are present and the dependencies should also be present. The Bleeding Edge repository can also maintain various tools including Aircrack-ng, dnsrecon, sqlmap, and beef-xss. You should also note that it is impossible to move some of these tools from their respective repositories to the Debian repositories. The Bleeding Edge repository can be added to the sources. List using this command:

Dpkg

This is a package management system that is also based on Debian. It is possible to remove, query, and also installs different packages when you are using the command-line application. After triggering the dpkg-1, some data will be returned in the process. In the process, you can also view all the applications that have been pre-installed into the Kali Linux operating system. To access some of the applications, you should make use of the command line.

Using Advanced Packaging Tools

The Advanced Packing Tools (APT) is essential when you are extending the dpkg functionally when searching and installing the repositories. Some of the packages may also be upgraded. The APT comes in handy when a user wants to upgrade the whole distribution.

The common APT (Advanced Packaging Tools) is as follows:

Apt-Get Upgrade - This is a command that is used to install the latest versions of various packages that have been installed on Kali Linux. Some of these packages have also been installed on Kali Linux and it is also possible to upgrade them. If there are no packages present, you cannot upgrade anything. Only installed packages can be upgraded.

Apt-Get Update - This is a command that is used when resynchronizing the local packages with each of their sources. Ensure that you are using the update command when performing the upgrade.

Apt-Get Dist-Upgrade - The command upgrades all the packages that are already installed in the system. The packages that are obsolete should also be removed.

To view all the full descriptions of some of the packages, you should use the apt-get command. It is also possible to identify the dependencies of each package. You may also remove the packages using various commands. Also, it is good to note that some packages may also not be removed using the apt-get command. You should update some packages manually using the update.sh script and you should also use the commands that are shown below:

Customizing and Configuring Kali Linux

The Kali Linux operating system framework is quite useful when performing penetration tests. As a security expert, you will not be limited to using the tools that have been pre-installed in Kali Linux. It is also possible to adjust the default desktop on Kali Linux. After customizing Kali Linux, you can also make sure that the system is more secure. After collecting some data, it will also be safe and the penetration test can also be carried out easily.

The common customizations include:

You can reset the root password.

You can add a non-root user.

Share some folders with other operating systems such as Microsoft Windows.

Creating folders that are encrypted.

Speeding up the operations at Kali Linux.

Resetting the Root Password

Use the following command so that you can change the root password:

Key in the new password. The following screenshot will guide you:

How to Add a Non-Root User

There are many applications that are provided by Kali Linux and they usually run as long as the user has the root-level privileges. The only issue is that the root-level privileges have some risks and they may include damaging some applications when you use the wrong commands when testing different systems. When testing a system, it is advisable to use user-level privileges. You may create a non-root user when using the adduser command. Start by keying in the following command in the terminal window. This screenshot will guide you:

How to Speed Up the Operations on Kali Linux

You can use different tools to speed up the processes in Kali Linux:

When creating the virtual machine, ensure that the disk size is fixed and that way it will be faster as compared to a disk that is allocated in a dynamic manner. As for the fixed disk, it will be easy to add files fast and the fragmentation will be less.

When using the virtual machine, make sure that you have installed the VMware tools.

To delete the cookies and free up some space on the hard disk, use the BleachBit application. To ensure that there is more privacy, ensure that the cache has been cleared as well as the browsing history. There are some advanced features such as shredding files and also wiping the disk space that is free. There are some traces that cannot also be fully deleted since they are hidden.

Preload applications exist and they can also be used to identify different programs that are also used commonly by various users. Using these applications, you can preload the binaries and the dependencies onto the memory and that will ensure that there is faster access. Such an application will also work automatically after ensuring that the installation process is complete.

Although Kali Linux has many tools, they are not all present on the start-up menu. The system data will also slow down when an application has been installed during the start-up process and the memory use shall also be impacted. The unnecessary services and applications should also be disabled; to do so, make sure you have installed the Boot up Manager (BUM). This screenshot will guide you:

You can also launch a variety of applications directly from the keyboard and make sure that you have added gnome-do so that you can access different applications from the accessories menu. After that, you will launch gnome-do and select the preferences menu and also activate the Quite Launch function afterward. Select the launch command and then clear all existing commands and enter the command line so that you can execute different commands after you have launched the selected keys.

Some of the applications can also be launched using various scripts.

Sharing the Folders with Another OS

Kali Linux has numerous tools. The operating system is also suitable since it offers some flexibility with regard to the applications that have already been pre-installed. To access the data that is present in Kali Linux and the host operating system, make sure that you are using the "Live Mode." You will then create a folder that you can also access easily.

The important data will be saved in that folder and you will then access it from either of the operating systems. The following steps will guide you on how to create a folder:

1) Create the folder on the operating system. For example, will be issued in the form of a screenshot, the folder, in this case, is named "Kali."

2) Right-click the "Kali" folder. You will then click 'share.'

3) Ensure that the file is shared with 'everyone'. People can also read and write an

4) My content is present in the "Kali" folder.

5)   You can also install some VMware tools of you have not yet shared and created the folder.

6)   After the installation process is complete, select the virtual machine setting. It will be present in the VMware menu. You will then share the folders and make sure that you have selected enabled. You will then create a path that allows you to select shared folders that are located in the primary operating system.

7)   Open the browser that is present on the Kali Linux default desktop. The shared folder will be present in the mint folder.

8)   Ensure that the folder has been dragged to the Kali Linux desktop.

9)   Make sure that all the information that has been placed into the folder is also accessible from the main operating system and Kali Linux.

When undertaking a pen test, make sure that you have stored all the findings in the shared folder. The information that you have gathered may be sensitive and you must ensure that it is encrypted. You can encrypt the information in different ways. For example, you can use LVM encryption. You can encrypt a folder or even an entire partition on the hard disk. Make sure that you can remember the password since you will not be able to reset it in case your memory fails you. If you fail to remember the password, the data will be lost in the process. It is good to encrypt the folders so that the data may not be accessed by unauthorized individuals.

Managing Third-Party Applications

Kali Linux has many applications and they are normally pre-installed. You may also install other applications on the platform but you need to make sure that they are from verifiable sources. Since Kali Linux is meant for penetration testing, some of the tools that are present on the platform are quite advanced. Before using these applications, make sure that you understand them fully so that you can use them effectively. You can also locate different applications easily.

Installing Third-Party Applications

There are many techniques that you can use when installing third-party applications. The commonly used techniques are such as the use of apt-get command and it is useful when accessing different repositories including GitHub and also installing different applications directly.

When you install different applications, make sure that they are all present in the Kali Linux repository. Use the apt-get install command during the installation process. The commands should be executed in the terminal window. During the installation process, you will also realize that the graphical package management tools will come in handy.

You can install different third-party applications and some of them include:

Gnome-Tweak-Tool - This is a tool that normally allows the users to configure some desktop options and the user can also change the themes easily. The desktop screen recorder will also allow you to record different activities that may be taking place on the desktop.

Apt-File - This is a command that is used to search for different packages that may be present within the APT packaging system. When using this command, you can list the contents of different packages prior to installing them.

Scrub - The tool is used to delete data securely and it also complies with various government standards.

Open office - This application offers users the productivity suite that will be useful during the documentation.

Team Viewer - This tool ensures that people can have remote access. The penetration testers can use the tool to carry out the penetration test from a remote location.

Shutter - Using this tool, you can take screenshots on the Kali Linux platform.

Terminator - The tool allows users to scroll horizontally.

There are numerous tools that are not available in the Debian repository and they can also be accessed using various commands such as apt-get install command which can also be installed on the Kali Linux platform. Users should first learn that the manual installation techniques involve the use of repositories and it is also possible to break down the dependencies which means that some of the applications may also fail in the process.

The GitHub repository has many tools and they are used mainly by the software developers when handling different projects. Some of the developers will prefer to utilize open repositories since they will gain a lot of flexibility. Different applications should also be installed manually. Make sure that you have also perused through the README file since it provides some guidance on how to use some of these tools.

Running the Third-Party Applications Using Non-Root Privileges

Kali Linux supports different activities such as penetration testing. There are some tools that can only run when a user has root-level access. Some of the data and tools may also be [protected using password and different encryption techniques. There are some tools that you can also run using the non-root privileges. Some of these tools are such as web browsers.

After compromising tools such as web browsers, the attackers will have some root privileges. To run applications as a non-root user, you should log in to Kali Linux first while using the root account. Ensure that the Kali Linux has been configured using a non-root account after that. An example will be provided whereby a non-root user account was formulated using the add user command.

The steps that you should follow are outlined below. In this instance, we will run the

Iceweasel browser and we will use the non-root Kali Linux user account.

Create a non-root user account.

We will use the sux application. The application is used when transferring different credentials from the root user account to the non-root user. When installing the application, use the apt-get install command.

You can launch the web browser and you should minimize it after that.

Use this command: ps aux | grep Iceweasel. In this case, you will be running the browser using the root privileges.

Close the browser and then start it all over again. Use the sux- noroot Iceweasel command to relaunch the application. The screenshot below will offer some subtle guidance.

Examine the browser title bar and you will realize that the browser was run as a non-root user and no administrator privileges are present.

Observe all the open processes after ensuring that the browser is running under the noroot account.

Effectively Managing the Penetration Tests

When you perform the penetration tests, you will come across a series of challenges and every test will also be carried out to unveil different vulnerabilities that may be present within the network or server. In some instances, you may not remember that you had conducted some tests and you may also be unable to keep track of the tests that you had already completed.

Some of the penetration tests are quite complex and the methodology used must adapt to that of the target. There are many applications that may be used when performing the tests and they include keyloggers and also Wireshark, just to mention a few. Each application is used when performing specific tests. The data that is gathered using these applications comes in handy. After the packets have been analyzed, it is easy to identify the packet tools that may have been affected.

There are many tools that are present in Kali Linux and some of them can also be used to make some rapid notes while serving as repositories using the KeepNote desktop wiki and Zim. Testers will also be able to carry out a variety of tests. In the process, they will collect some data that will also be used to facilitate the tests. The tests help to identify some of the changes that have taken place in the system. Some vulnerabilities may emerge and they should be sealed immediately to ensure that external attackers will not be able to access the system and gain access to sensitive pieces of information.

Chapter 2: Back Door Attacks

Imagine you're going to a concert, but you don't have a ticket. You see the line of people all with their purchased tickets waiting to get through security. You see cameras pointing at the front door and a few extra security guards guarding the sides. You don't have a ticket or the money to buy one. Then, you see a little unguarded, dark, hidden alley with no cameras and the back door. The doors that lead to the venue. They are unlocked, and there are no security or cameras around. Would you go through the door? That's the concept behind a back-door attack.

How do backdoors even end up on our computers? Well, they can end up there intentionally by the manufacturer; this is built in so they can easily test out the bugs and quickly move in the applications as they are being tested.

The back door can also be built by malware. The classic backdoor malware is the infamous

Trojan. Trojan subtly sneaks up on our computer and opens the back door for the people using the malware. The malware can be hidden into anything—a free file converter, a PDF file, a torrent, or anything you are downloading into your computer. Of course, the chances are higher when what you're downloading is a free copy of an otherwise paid product (lesson to be learned here). Trojans have an ability to replicate, and before you know it, your computer is infected with malware that is opening a backdoor for the whole line up to come in to see the show for free.

The back door can be used to infiltrate your system not only for passwords but also for spying, ransomware, and all kinds of other malicious hacking.

How to protect yourself from back door attacks

Choose applications, downloads and plugins carefully; free apps and plugins are a fantastic thing, but YouTube to MP3 converters, torrents of the latest Game of Thrones season, and a copy of Photoshop might not be the best option if you're interested in keeping your passwords safe. Android users should stick to Google Play apps, and Mac users should stick with the Apple store. Track app permissions too and be sure to read, at least a little, before you sign your life away to a third-grade flashlight application.

You can also try:

Monitoring network activity - Use firewalls and track your data usage. Any data usage spike is a sure sign of backdoor activity.

Changing your default passwords - When a website assigns a default password, we may find that we are just too lazy to take the 30 seconds necessary to change it. Just do it. You might not be locking the back door with the latest state-of-the-art security system, but at least you are not keeping them wide open with a neon sign pointing to your password. Freckles might be your puppy, but he can't be a password for everything. A common complaint is, "I will forget it." Write it down. Contrary to popular belief, hackers won't go into your house and search for that piece of paper, but they will go into your computer. Which option seems safer?

Zombie Computers for Distributed Denial of Service (DDoS) attacks

Sounds extremely cool, right? Well, it's not. Basically, a computer becomes a zombie computer when a hacker infiltrates it and controls it to do illegal activities. The best part (for the hacker, not for you) is that you are completely unaware that all this is happening. You will still use it normally, though it might significantly slow down. And then all of a sudden, your computer will begin to send out massive spam emails or social media posts that you have nothing to do with. DDoS attacks are lovely (for the hacker, not for you) because they work on multiple computers at once, and the numbers can go into millions. A million zombie computers are mindlessly wandering around the internet spamming everything in sight, infecting other computers. The version where your computer is infected only to send out spam is the light version. DDoS attacks can also be used for criminal activity, and this is why it is important to prevent them.

How to protect from DDoS attacks

Larger scale businesses require more substantial protection against DDoS attacks, and we will go over that in detail, but even for individuals, half of the protection is prevention.

Understand the warning signs—slowed down computers, spotty connection, or website shutdowns are all signs of a DDoS attack taking place.

What can you do?

Have more bandwidth - This ensures you have enough bandwidth to deal with massive spikes in traffic that can be caused by malicious activity

Use anti-DDoS hardware and software modules - Protect your servers with network and web application firewalls. Hardware vendors can add software protection by monitoring how incomplete connections and specific software modules can be added to the webserver software to provide DDoS protection.

Smart clicking - This should go without saying, but for those in need of hearing it—pop-up ads with a "No, thanks" button are hateful little things. Just exit the website, don't click anything on that ad, especially not the "No, thanks," button or you will instantly activate an annoying download, and now your computer is a zombie.

Man in The Middle

When you're online, your computer does little back-and-forth transactions. You click a link, and your computer lets the servers around the world know you are requesting access to this website. The servers then receive the message and grant you access to the requested website. This all happens in nanoseconds, and we don't think much about it. That nanosecond moment between your computer and the web server is given a session ID that is unique and private to your computer and the webserver. However, a hacker can hijack this session and pretend to be the computer and as such, gain access to usernames and passwords. He becomes the man in the middle hijacking your sessions for information.

How to protect yourself from the man in the middle

Efficient antivirus and up-to-date software go a long way in preventing hijacking, but there are a couple of other tips that can help you prevent becoming a victim.

Use a virtual private network - A VPN is a private, encrypted network that acts as a private tunnel and severely limits the hacker's access to your information. Express VPN can also mask your location, allowing you to surf the web anonymously wherever you are.

Firewalls and penetration testing tools - Secure your network with active firewalls and penetration testing tools.

Plugins - Use only trusted plugins from credible sources and with good ratings.

Secure your communications - Use two-step verification programs and alerts when someone signs in to your account from a different computer.

Root Access

Root access is an authorization to access any command specific to Unix, Linux, and Linux-like systems. This gives the hacker complete control over the system. Root access is granted with a well-designed rootkit software. A quality designed rootkit software will access everything and hide traces of any presence. This is possible in all Unix-like systems because they are designed with a tree-like structure in which all the units branch off into one root.

The original Unix operating system was designed in a time before the personal computer existed when all the computers were connected to one mainframe computer through very simple terminals. It was necessary to have one large, strong mainframe for separating and protecting files while the users simultaneously used the system.

Hackers obtain root access by gaining privileged access with a rootkit. Access can be granted through passwords; password protection is a significant component in restricting unwanted root access. The rootkit can also be installed automatically through a malicious download. Dealing with rootkit can be difficult and expensive, so it's better to stay protected and keep the possibility of root access attacks to the minimum.

How to protect yourself from root access attacks

Quality antivirus software is one of the standard things recommended in all computers, be it for individuals or businesses. Quality antivirus helps the system hardening making it harder for installation of rootkits.

Principle of least privilege - PoLP gives only the bare minimum privilege that a program needs to perform allows for better protection from possible attacks. For example, in a business, a user whose only job is to answer emails should only be given access to the emails. If there is an attack on the user's computer, it can't spread far because the person only has access to email. If a said employee has root access privilege, the attack will spread system-wide.

Disable root login - Servers on most Unix and Linux operating systems come with an option for the root login. Using root login allows for much easier root access, and if you pair it with a weak password, you are walking on a thin line. Disabling the option for root access keeps all the users away from the root login temptation.

Block brute forces - Some programs will block suspicious IP addresses for you. They will detect malicious IP and prevent attacks. While manually detecting is the safest way, it can be a long process; programs that are designed to block malicious IPs can drastically save time and help prevent root access attacks.

The best way to protect yourself from hack attacks is through prevention because the alternative can be lengthy, exhausting, and costly.

Chapter 3: Cybersecurity

The internet is a vast place, and most people are not experts on protecting the information about them that is available. It's no surprise that there are people out there who take advantage of others' ignorance. But there are ways to protect yourself from those kinds of attacks, and that's where cybersecurity comes in.

What is Cybersecurity?

By the time you finish reading this sentence, over 300 million people will have clicked on a single link. You are part of a universe that generates information every millisecond. We do everything from home—buy, sell, eat, drink, fight, tweet, click, and share. We don't need to go to the movies to see a movie or go to the stores to shop. Information exchanges happen online every time you connect to Wi-Fi, publish content, buy something online, like a post on social media, click a link, send an email...you get the gist. We produce much more information than we can grasp, so we underestimate the quantity and value of protecting it.

Cybersecurity is the protection of hardware, software, and data from cyberattacks. Cybersecurity ensures data confidentiality, availability, and integrity. A successful and secure system has multiple layers of protection spread across the networks, computers, data, and programs. For cybersecurity to be effective, all the people involved in different components must complement each other. It is always better to prevent cyberattacks then deal with the consequences of one.

Cyberattacks hit businesses every day. The latest statistics show that hackers now focus more on quieter attacks, but they are increased by over 50%.

During 2018, 1% of websites were considered victims of cyberattacks. Thinking about 1% of all websites that exist, that adds up to over 17 million websites that are always under attack. Cyberattacks cost an average of $11 million per year, so cybersecurity is a crucial aspect of saving your business much money.

That's where the most prominent problem occurs. Small business owners and individuals don't grasp the potential threat to their data because they don't see the value they bring to a hacker attacking. The value is in the lack of security.

Many small businesses with no security are more accessible to penetrate than one large corporation. Corporations invest in cybersecurity; small business owners and individuals do not. They use things like the cloud. Their data migrate with them to the cloud allowing criminals to shift and adapt. The lack of security on their part is crucial to these statistics. The most definite form of on-going attacks remains ransomware; it is so common-place that it is barely even mentioned in the media. Ransomware infects a website by blocking access to their data until a business or an individual transfer a certain amount of money. Hackers hold your data hostage, and it's always about the money.

Cybersecurity is not complicated, it is complex. However, it is also very important to understand. Implementing just the top four cybersecurity strategies diminishes attacks by over 70%. Here are some of the techniques:

Application whitelisting - allowing only approved programs to run

Applications security patching - enforcing security patches (fixes) promptly for applications

Operating systems security patching - enforcing security patches (fixes) promptly for the whole system

Limiting administrative privileges - allowing only trusted individuals to manage and monitor computer systems

Cybersecurity Benefits

There's a variety of benefits cybersecurity can bring to you or your business, and some aren't as obvious as you may think.

Prevents ransomware - Every 10 seconds, someone becomes a victim of ransomware. If you don't know what is happening in your network, an attacker probably found a way to get into it.

Prevents adware - Adware fills your computer with ads and allows the attacker to get into your network.

Prevents spyware - The attacker can spy on your activity and use that information to learn about your computer and network vulnerabilities

Improves your search engine rankings - SEO is the key in the modern digital market. Small businesses looking to rank up on search engines have to be educated in SEO if they want to advance financially. HTTPS (HyperText Transfer Protocol Secure), or the encryption of username, passwords, and information, is one of the critical SEO ranking factors.

Prevents financial loss and saves your startup - More than half of small business go down after a cyberattack. The downtime required to fix the damage prevents any new business, and the data breach causes you to lose the trust of your current customers. Stable businesses can find a way to recover from this, but startups rarely make it out alive.

Chapter 4: Wireless Networking

While talking about networking, one of the most trending topics is wireless networking. It has allowed people to reach new heights of reliability along with benefits which allow them to use the internet with their devices without any form of cable or wire in between. All of these have been possible only because of wireless networking. In wireless networking, all the devices connect with a network switch or router which helps in establishing connection between the devices and the Web via radio waves. All the information and connection are established through the air. Thus, it can be regarded as a mobile form of network where you are no longer required to be seated in one single place for surfing the internet. Wireless networking comes with various components along with some very interesting features which will be discussed further in this chapter. So, let's start with wireless networking and its various features.

Hacking and Penetration Testing with Kali Linux

Each and every organization and companies come with certain weak points which might turn out to be malicious for the organization. Such weak points can also lead to some serious form of attack which can be later used for manipulation of organizational data. The only thing that you are left with in such a situation which can ultimately help you in preventing all forms of hackers from getting into your systems is regular checking of infrastructure security. You will also need to ensure that no form of vulnerability is present within the infrastructure. For serving all of these functions, penetration testing is something which can ultimately help you. It helps in detecting the vulnerabilities within a system and forwards the same information to the organization administrators for mending up the gaps. Penetration testing is always performed within a highly secure and real form of environment which helps in finding out the real form of vulnerabilities and then mends the following along with securing the system.

Details about penetration testing

It is a process which is used for testing of the systems for finding out or ensuring that whether any third party can penetrate within the system or not. Ethical hacking is often being mixed up with penetration testing as both of them somewhat serves the similar purpose and also functions more or less in the same way. In penetration testing, the pen tester scans the systems for any form of system vulnerability, flaws, risks and malicious content. You can perform penetration testing either in an online form of environment or server or even in a computer system. Penetration testing comes with some ultimate form of agendas: strengthening the system of security and defending the structure of an organization from potential attacks and threats.

Penetration testing is absolutely of legal nature and is done along with the other official workings. When used in the proper and perfect way, penetration testing has the ability of doing wonders. If you want you can also consider penetration testing as a potential part of ethical hacking. You will need to perform the penetration tests at a regular form of intervals as it has the ability of improving the system capabilities. It also helps in improving the cyber security strategies. In order to fish out all the weak points within a program, system or application, various forms of malicious content are constructed or created by the pen testers. For an effective form of testing, the harmful form of content is spread across the overall network for the testing of vulnerability.

The technique used by penetration testing might not be successful in handling all the security concerns but it can help to minimize the chances of probable attacks on the system. Penetration testing ensures that an organization or company is absolutely safe from all forms of threats and vulnerabilities and it ultimately helps in providing security from the cyber form of attacks. It also makes sure that the system of defense of an organization is working properly and is also enough for the company or organization to prevent the probable attacks and threats. Not only that but it also indicates the measures of security which are required to be changed by the organization for the only motive of defending the system from attacks and vulnerabilities. All the reports regarding penetration testing are handed over to the system administrators.

Metasploit

Metasploit is nothing but a framework meant for penetration testing which actually makes the concept of hacking much simpler and easier. It is regarded as an important form of tool for majority of the attackers along with the security defenders. All you need to do is to just point out Metasploit at the target, pick any exploit of your choice, choose the payload which you want to drop and just hit enter. However, it is not that casual in nature and so you will need to start from the beginning. Back in the golden days, the concept of penetration testing came with lots of repetitive form of labor which is now being automated by the use of Metasploit.

What are the things that you need? Gathering of information or gaining of access or maintaining the levels of persistence or evading all forms f detection? Metasploit can be regarded as the Swiss knife for the hackers and if you want to opt for information security as your future career then you are required to know this framework in detail. The core of the Metasploit framework is free in nature and also comes pre-installed with the software Kali Linux.

How to use Metasploit?

Metasploit can seamlessly integrate itself with SNMP scanning, Nmap and enumeration of Windows patch along with others. It also comes with a bridge to the Tenable's scanner of vulnerability along with Nessus. Most of the reconnaissance tools which you can think of can integrate along with Metasploit and thus it makes it possible to find the strongest possible point in the shield of security. After you have identified the weakness in a system, you can start hunting across the huge and extensible form of database for the need of the exploit which will help in cracking the strongest armor and will let you in the system. Just like the combination of cheese and wine, you can also pair an exploit with the payload for suiting any task at the hand.

Most of the hackers are looking out for a shell, a proper payload at the time of attacking a system based on Windows acts as the Meterpreter and also as an in-memory form of interactive shell. Linux comes with its own set of shellcodes which depends on the exploit which is being used. Once within a target machine, the quiver of Metasploit comes with a complete suite of post-exploitation tools which also includes escalation of privileges, pass the hash, screen capture, packet sniffing, pivoting and keylogger tools. If you want you can also easily set up a proper form of backdoor if the target machine gets rebooted somehow.

Metasploit is being loaded up with more and more features each year along with a fuzzer for identifying the potential flaws of security in the binaries as well as a too long list of the modules which are of auxiliary nature. What we have discussed till now is only a high-level vision of what can be done with Metasploit. The overall framework is modular in nature and can be extended easily and it also enjoys an active form of community. In case it is not doing what you want, you can easily tweak the same for meeting your needs.

How can you learn Metasploit?

You can find out various cheap as well as free forms of resources for the purpose of learning Metasploit. The best way of starting with Metasploit is by downloading Kali Linux followed by the installation of the same along with a virtual machine for practicing of the target. The organization which maintains Kali Linux and also runs the OSCP certification, Offensive Security, offers a free course that includes training of Metasploit and is known as Metasploit Unleashed.

Where can you download Metasploit?

Metasploit can be found along with the hacking software Kali Linux. But, if you want you can also download it separately from the official website of Metasploit. Metasploit can be used on the systems which are based on Windows and *nix. You can find out the source code of Metasploit Framework on GitHub. Metasploit is also available in various forms which you can easily find over the internet.

Datastore

The datastore can be regarded as a core element of the Metasploit Framework. It is nothing but a table of several named values which allows the users to easily configure the component behavior within Metasploit. The datastore allows the interfaces to configure any of the settings, exploits for defining the parameters and also payloads for the purpose of patching the opcodes. It also allows Metasploit Framework to pass internally between the options of modules. You can find two types of datastores, the Global datstore which can be defined by using 'setg' and the Module datastore which can be defined at the modular level of datastore by using 'set'.

SQL Injection and Wi-Fi Hacking

When it comes to cyber-attacks one of the most widely used forms of attack is the SQL Injection attack. In this, an attacker performs the attack by executing threat or invalid form of SQL statements which is used for database server control for an application of web. It is also being used for modifying, deleting or adding up records within the database without even the user knowing anything about the same. This ultimately compromises the integrity of the data. The most important step which can be taken for avoiding or preventing SQL injection is by input validation.

SQL Injection and its types

There are various types of SQL injection which you can find today. Let's have a look at them.

● Classic or In-band SQL injection: 1. Error based: Attackers employ the generated error by the database to attack the database server.

2. Union based: In this UNION SQL operator is employed for combining a response for returning to the HTTP response.

● Inferential or Blind SQL injection: 1. Based on Boolean: It is based on return of true or false.

2. Time based: It sends out SQL injection which forces the database just before responding.

- Out of band SQL injection: This takes place when an attacker is unable to use the similar form of channel for attacking and gathering the results.

Tools used for SQL injection

There are various tools which are being used for carrying out SQL injection.

- SQLMap: This tool is used for an automatic form of SQL injection and it is a tool which helps in taking over the database.

- jSQL Injection: It is a Java based tool which is being used for SQL injection.

- Blind-SQL-BitShifting: It is a tool which is used for blind SQL injection by the use of BitShifting.

- BBQSQL: It is a blind form of SQL injection exploitation tool.

- explo: It is a format of machine and human readable web vulnerability testing.

- Whitewidow: It is a scanning tool which is used for checking out the vulnerabilities of the SQL database.

- Leviathan: It acts as an audio toolkit.

- Blisqy: It is used for the purpose of exploiting time-based SQL injection within the headers of HTTP.

Detection tools for SQL injection

A tool named Spider testing tool is widely being used for the purpose of identifying the holes of SQL injection manually by the use of POST or GET requests. If you can resolve the vulnerabilities within the code then you can easily prevent the SQL injections. You can also take help of a web vulnerability scanner for identifying all the defects within the code and for fixing the same f0r preventing SQL injection. The firewalls present in the web application or within the application layer can also be used extensively for preventing any form of intrusion.

Hacking of Wi-Fi

Wi-Fi or wireless networking can be regarded as the most preferred medium which is being used for the purpose of network connectivity in today's world. However, because of so much popularity of the same, the wireless networks are also subjected to various attacks and also comes with several issues of security. In case the attacker gains complete access of the network connection then it is possible for the attacker to sniff off the data packets from any nearby location. The attackers employ sniffing tools for finding out the SSIDs and then hacks the Wi-Fi or wireless networks. After successful hacking, the attackers can monitor all the devices which are connected with the same SSID of the network. In case you use authentication of WEP then it might be subject to dictionary attack. The attackers employ RC4 form of encryption algorithm for the purpose of creating stream ciphers which is very easy to be cracked. In case you are using authentication of WPA then it might subject to DOS along with dictionary attack.

Tools for hacking of Wi-Fi

For the purpose of cracking WEP, the attackers use various tools such as WEPcrack, Aircrack, Kismet, WEPDecrypt and many others. For cracking WPA, tools such as Cain, Abel and CowPatty are being used by the hackers. There are also various types of tools which are used in general for hacking of wireless network system like wireshark, Airsnort, Wifiphisher, Netstumbler and many others. Even the attackers are now able to hack the mobile phone platform via the wireless network system. Android can be regarded as the most found mobile phone-based platform but it is also very much susceptible to some specific types of vulnerabilities which ultimately makes it easier for the attackers to exploit the device security and then steal data from the same. The most dangerous threats for the mobile devices are third party applications, email Trojans, wireless hacking and SMS.

How are Wi-Fi attacks carried out?

Most of the wireless network attacks are carries out by setting up rogue Access Point.

• Evil Twin attack: In this, the hacker sets up a false access point with the same name as that of the corporate AP which is close to the premises of the company. When any employee of the company connects to that access point by regarding that access point to be genuine in nature, that employee unknowingly gives out all the details of authentication of the actual access point. Thus, the hacker can easily compromise the overall connection.

• Signal jamming: The hackers can easily disrupt the network connection which can be done by jamming the network signals. This is done by various forms of tools which are used for creating noise.

• Misconfiguration attack: When the router of a network is set up by using a default form of configuration, weak form of encryption, weak credentials and algorithms, an attacker can easily crack the network.

● Honey spot attack: The attackers set up false hotspots or access points with the same name of the SSID similar to any public Wi-Fi access point. When any user connects with that access point unknowingly, the hackers can easily get access to the actual network.

How To Carry Out An Effective Attack

When it comes to the term 'hacking' it doesn't mean that it has to be negative all the time. You will be able to have a proper idea about the overall process of hacking only when you will have a clear perception about the process behind it. Not only that you will be able to gather knowledge about the process of hacking but you will also be able to make your system much more protected from external attacks. Most of the times, when an attacker tries to gain access to a server of an organization or a company, it is generally done by using 5 proper steps. Let's have a look at those steps.

●   Reconnaissance: This can be regarded as the very first step that comes in the hacking process. During this phase, the attacker uses all the available means for the purpose of collecting all forms of relevant information about the primary target system. The relevant set of information might include the proper identification of the target, DNS records of the server, range of the IP address which is in target, the network and various other aspects. In simple terms, the attacker tries to collect all sorts of information along with the contacts of a website or server. This can be done by the attacker by the use of several forms of search engines like maltego or by researching about the system which is in target or by using the various tools like HTTPTrack for the purpose of downloading a complete website for enumeration at a later stage.

By performing all these steps, the attacker will be able to determine the names of all the staffs within an organization very easily, find out the designated posts along with the email addresses of the employees.

• Scanning: After collecting all the relevant information about the target, the attacker will now start with the process of scanning. During this phase, the attacker employs various forms of tools like dialers, port scanners, vulnerability scanners, sweepers and network mappers for the sole purpose of scanning the target website or server data. During this step, the attackers try to seek out all that information which can actually help in the execution of a successful attack such as the IP address of the system, the user accounts and the computer names within that server. Right after the hackers are done with scanning of basic information, they start to test the network which is in target for finding out the possible avenues of attack. They might also employ several methods for network mapping just like Kali Linux.

The hackers also search out for any automatic email system by which they can mail out the staffs of the target company about some false form of query like mailing the company HR about a job query.

• Access gaining: This is the most important of all the steps. In this phase, the attacker designs out the blueprint of the target network along with the help of all relevant information which is also collected in the first and second step of hacking. As the hackers are done with enumeration of data followed by scanning of the system, they will now move to the step of gaining access to the system which will be based on the collected information.

For instance, the attacker might decide to use a phishing attack. The attackers will always try to play safe and might employ a very simple attack of phishing for gaining overall access to the system. The attacker might also penetrate into the system from the IT department shell. The hackers use phishing email by employing the actual email address of the company. By using this phishing email ID, the attacker will send out various emails to the techs that will also contain some form of specialized program along with a phishing website for gathering information about the login passwords and IDs. For this, the attackers can use various methods such as phone app, website mail or something else and then asking the employees to login with their credentials into a new website.

As the hackers use this method, they already have a special type of program running in the background which is also called as Social Engineering Toolkit which is used for sending out emails with the address of the server to the users.

● Maintaining access to the server: After the attackers have gained access to the target server, they will try out every possible means for keeping their access to the server safe for future attacks and for the purpose of exploitation. As the attacker now has overall access to the server, he might also use the server as his very own base for launching out several other forms of attacks. When an attacker gains access to an overall system and also owns the system, such a system is called as zombie system. The hacker might also try to hide himself within the server by creating a new administrator account with which he can easily mingle with the system without anyone knowing about it. For keeping safe access to the system, the hacker traces out all those accounts which are not being used for a long time and then elevates the privileges of all those accounts to himself.

As the hacker makes sure that no one has sensed his presence within the system, he starts to make copies of all the data on that server along with the contacts, messages, confidential files and many more for future use.

● Clearance of tracks: Right before starting with the attack, the hackers chalk out their entire track regarding the identity so that it is not possible for anyone to track them. The attackers begin by altering the system MAC address and then run their entire system via a VPN so that no one can trace their actual identity.

Chapter 5: How to Initiate A Hack Using Kali Linux?

When planning an attack, the most important factor to consider is the pilot study. It should come first before you carry out an attack or a penetration test on a target. As an attacker, you will have to dedicate a lot of time to the reconnaissance. In this stage, the attacker will be able to define, map, and also explore some of the vulnerabilities that are present and they will be able to successfully perform an exploit. There are two types of pilot studies; passive and active.

The passive pilot study involves the analysis of the information that is available. For instance, some information can be obtained online through search engines. The information can be analyzed first. Although an attacker can use this information to their advantage, it is not possible to trace the information back to them. As for passive reconnaissance, it is mainly carried out to ensure that the target cannot easily notice that there is a looming attack.

The major practices and principles of the passive reconnaissance include:

OSINT (Open-source intelligence).

How to obtain user information.

The basics of the pilot study.

The Basic Principles of the Pilot Study

The pilot study is the first step when a person wants to launch an attack. The study is carried out after identifying a target. The information that is gained during this stage will come in handy when performing the actual attack. A reconnaissance will ensure that they have provided a sense of direction which will be required when trying to look into some of the vulnerabilities that are present in the network or target's server.

The passive pilot study does not involve physically interacting with the target and that means that the IP address of the attacker is not logged. For instance, the attacker may search for the IP address of the target. It may be difficult to gain access to such information; however, it is also possible. The target will also not be able to notice that an attacker is trying to harvest some information as they plan an attack.

The passive reconnaissance will focus more on the business activities as well as the employees within the organization. The information that is readily available on the internet is known as OSINT (Open source Intelligence).

As for the passive reconnaissance, the attacker will interact with the target in a manner that is expected. For instance, the attacker will visit the website of the attacker. They will then view the available pages and they will then download some of the available documents. Some of these interactions are always expected and they are not detected easily and the target may not know that there is a looming attack.

The active reconnaissance involves interacting through port scanning in the specific network as well as sending direct queries that will then trigger the system alarms and that means that the target can easily capture the IP address of the attacker and their activities. The information that the target has gained can also be used to arrest the attacker. Additionally, the information can also be presented before a court of law as evidence that the attacker was planning something malicious. As for the active pilot study, there are various activities that the attacker should consider so that they can conceal their identity.

As an attacker, you should also follow some steps during the process of gathering information. The main focus is on the user account data. For the pilot study to be effective, as an attacker, you should always know what you are looking for. Also, make sure that you have gathered all the data that you need. Although the passive reconnaissance is less risky, it minimizes the amount of data that you can collect.

OSINT (Open-Source Intelligence)

This is the first step when planning an attack. In this case, the attacker should make use of the present search engines preferably Google. There is a lot of information that could come in handy when facilitating an attack. The process of collecting the information is quite complex.

In this book, we will just issue an overview since the main focus is on how to hack with Kali Linux. The essential highlights will offer some suitable guidance. The information collected by an attacker will always depend on their initial motives and their major goals when they plan an attack. For instance, the attacker may want to access the financial data within a specific organization. Other types of information that they may need is the names of the employees. Most of the attackers will focus more on the senior employees who are working as executives. Some of these employees include the CFO among other seniors. The attacker will focus on obtaining their usernames and their respective passwords. In some instances, an attacker may try to carry out social engineering. In this case, they will have to supplement the information that they possess so that they may appear as credible individuals. After that, they can easily request for the information that they need.

As for the Open source Intelligence, the attacker will start by reviewing the online presence of the target. They will start by observing their social media pages, blogs, and websites. The public financial records also come in handy in some cases. The most important information is:

The geographical location of the offices. For instance, there can be some satellite offices that also share some corporate information but they have not set up any measures that will ensure that the information is safe as it is being transmitted from one office to another.

The overview of the parent and subsidiary firms matters especially when dealing with a new company that has also been acquired through M&A transactions. The acquired companies will not be as safe as compared to the parent company.

The contact information and the names of the employees. The phone numbers and email addresses should also be obtained.

Looking for clues about the target company's corporate culture so that it may be possible to facilitate the social engineering attack.

The business partners are also eligible to access to network of the target.

The technology being used. For instance, the target may issue a press release about how to adopt software and the attacker will go ahead and review the website of the vendor as they try to look for bug reports. After finding some vulnerabilities, they will be able to launch an attack.

Some of the online information sources that can also be used by an attacker when they are planning an attack include:

Search engines including Google. There are also other search engines such as Bing. It's only that we have gotten used to Google. During the search process, you will realize that the process is highly manual. You may have to type the name of the company as well as other relevant details. Since technology has also advanced, there are some APIs that can be used to automate the searches of the search engines. Some of the effective APIs include Maltego.

There are other sources and they include:

The financial and government sites since they provide some information about the key individuals within the company as well as some supporting data.

The Usenet newsgroups. The man focus should be on the posts by the employees that you are targeting as a tester or an attacker.

You may also seek some help with different forms of technology.

Jigsaw and LinkedIn; these companies come in handy since they provide some information about the employees within a company.

The cached content. It can be retrieved easily by search engines including Google.

The country as well as the specific language being used.

Employee and corporate blogs.

Social media platforms such as Facebook.

The sites whereby you can look up the server information and the DNS as well as routes. Some of these sites include myIPneighbors.com.

The main issue arises when you have to manage the information that you have found. The main advantage is that kali Linux has an application known as Keep Note. It supports the rapid importation and management of different data types.

Route Mapping DNS reconnaissance

As a tester or an attacker, you will have to make sure that you have identified the targets that have an online presence. Make sure that you have also gained access to some of the items that may pose some interest. You will then go ahead and identify the IP addresses of the targets. The DNS reconnaissance will come in handy when identifying the domains as well as the DNS information that will help to define some of the IP addresses as well as actual domain names. The route between the attacker and the target will also be identified.

The information is easily available in some of the open sources. Some information is mainly present in some of the DNS registrars and they are referred to as third parties. The registrar may collect an IP address as well as some of the data requests that have been brought forth by an attacker. Such information is rarely provided to the specific target who will be a victim of an attack. As for the target, they can easily monitor the DNS server logs. The information needed can also be obtained using an approach that is systematic.

WHOIS

The first step entails researching the IP address so as to identify the addresses that have also been assigned to the sites of the target. You will then make use of the whois command and it will allow you to query the databases that have also stored the information about certain users. The information that you will obtain includes the IP address and domain name.

The whois request will then come in handyu when providing physical addresses, names, e-mail addresses, as well as phone numbers. Such information is very important when it comes to performing a social engineering attack.

As an attacker or tester, you can use the whois command to carry out the following activities:

Supporting a social engineering attack against a target that has been identified using the whois query.

Identifying the location whereby you can launch a physical attack.

Conducting some research that will allow you to learn more about the domain names that are present on the server. You can also learn more about the number of users operating it. As an attacker, you will also gain an interest in learning whether the domains are insecure and whether you can exploit the present vulnerabilities to gain access while also compromising the target server.

Identifying the phone numbers since you may also have to launch a dialing attack while conducting the social engineering attack.

The attack will then use the DNS servers to carry out the DNS reconnaissance.

In some cases, the domain may be due to expire and the attacker may go ahead and try to seize the domain while also creating look-a-like website that will be used to lure unsuspecting visitors who think that they are entering into the original website.

To make sure that the data has been shielded accordingly, there has been an increase in the use of third parties. Also, when using public domains, you cannot access domains such as .gov and .mil. The mentioned domains belong to the military and the government and that is why they have been secured so that they cannot be accessed by other parties. When you send a request to such a domain, it will be logged. There are many online lists that can also be used to describe the IP addresses as well as domains. If you want to use the whois query, the following screenshot will offer some guidance when running the query against some of the Digital Defense domains:

There is a whois command record that will be returned and it will contain some names and geographical information as well as contact information that will come in handy when facilitating a social engineering attack. There are also many websites that are also used to automate the whois lookup. Some of the attackers use some of these sites to insert a step that will be between them and the attackers. The site that is doing the lookup may then log the IP address of the requester.

Mapping the Route to the Target

The route mapping was once used as a diagnostic tool. The tool would allow the attacker to view the route that is followed by the IP packet as it moves from one host to another. When using the TTL (time to live) field in the IP packer, an ICMP TIME_EXCEEDED message will then be elicited from one point to another. The message will be sent from the receiving router and it will also help to determine the value that is in the TTL field. The packets will also count the number of routes and hops that have been taken.

From the perspective of the attacker or penetration tester, the traceroute data will help to yield the following pieces of data:

The hints about the topology of the network.

The path that is present between the target and the attacker.

Identifying the firewalls and other devices that are used to control access to the network.

Identifying whether the network has been misconfigured.

In Kali Linux, you can map the route using the tracerouteis command. If you are using Windows, you can use the tracert command. If you happen to launch an attack when using Kali Linux, you will notice that most of the hops have been filtered. For instance, when using Google to trace the location of a certain target, the results will be as shown below:

If you were to run the same request when using the tracert on the Windows platform, you will see the following:

We will get the complete path and we have also noticed that Google is showcasing an IP address that is slightly different. The load balancers have also been indicated. The main reason why the path data is different is because the traceroute used the UDP datagrams whereas the Windows tracert will use the ICMP request (specifically the ICMP type 8). When you complete the traceroute when using the tools that have been provided by Kali Linux, you should also make sure that you have used multiple protocols so that you may obtain the complete path while also bypassing some of the devices that carry out packet-filtering.

Obtaining User Information

When an attacker or a penetration tester manages to gather the usernames and the e-mail addresses of the targets, they can then manage to gather into the systems. The most common tool that is deployed is the web browser and you have to perform a manual search. You have to search some of the third-party sites including Jigsaw and LinkedIn. You can also use some of the tools provided by Kali Linux to automate the search.

Chapter 6: The Hacking process & Kali Linux Installation

This chapter explains to us the hacking process that beginner hackers should master to get a good overview of hacking and its importance. Although being a little practical, this chapter will get you started and help you understand the basic things you need to know for becoming a professional hacker. We will also explain how to install a virtual machine and Kali Linux in this chapter. Let us start!

Essential things for a hacker

1) First, have a Basic English understanding:

Knowing English is critical for hackers, as most instructions made for them are now in English. Therefore, beginner hackers should try to read English materials, use English software, while paying attention to foreign network security at the same time. You may occasionally use foreign resources to master hacking methods and techniques.

2) Second, use and learn basic software:

The basic software cited here has two major components. One is the common computer commands we use every day, such as FTP, ping, net, etc., while the other is learning about primary hacking tools. Port and Vulnerability Scanners, Information Interception Tools, and password cracking tools. This software has many uses and functions. This book is going to introduce several popular software usage methods. After learning the basic principles, learners can choose either their own tools or create their own tools. Find the development guide for the software and write to make your signature hacking tools for a better understanding of the network structure.

3) Third, an elementary understanding of network protocol and working principle is a must:

The so-called "preliminary understanding" is to "get their own understanding on the topic" to understand the working principle of the network, because the knowledge involved in the agreement is complex, I mean very complex if you do in-depth research at the beginning, it is bound to/Will greatly dampen the enthusiasm for learning. Here I suggest that learners have a preliminary understanding of the TCP/IP protocol, especially how the network communicates and how information is exchanged when browsing the web, how the client browser applies for "handshake information," how the server "responses to handshake information" and "accepts requests."

4) Get to know several popular programming languages and scripts:

There is no requirement for learners to learn thoroughly, as long as you know the results of program executions. It is recommended that learners initially learn Python, ASP, and CGI scripting language, and have an elementary understanding of HTML, PHP, and Java, etc., you need to concentrate mainly on the "variables" and "array" parts of these languages because there is an inherent connection between languages. In a way, such that so long as you are proficient in one of them, other languages can come later. It is recommended to learn C language and HTML.

5) Get intimate with a web application:

Web applications include various servers' software daemons, such as wuftp, Apache, and other server backgrounds. There are various popular forums and e-communities on the Internet. Conditional learners should make their own computers into servers, and then install and run some forum code. After some test runs, they will be sensible to understand the working principle of the network, which is much easier than relying on theoretical learning. Try to do more with less work.

Some important concepts you need to master before hacking:

I. The Protocol

Networks are places where information is exchanged. All computers accessing the network can exchange information through a physical connection between devices. Physical equipment includes the most common cables, optical cables, wireless WAPs, and microwaves. However, merely possessing these physical devices does not enable information exchange. It is the same when the human body is not controlled by the brain, and the information exchange must have a software environment. This software environment is a set of rules that humans have implemented. It is called a protocol. With a protocol, different computers can use physical devices in accordance with the same protocol and do not cause mutual incomprehension.

This kind of agreement is very similar to Morse code. It can be changed in a simple way. However, if there is no control table, no one can understand what the content of a chaotic code is. The same is true for computers, which accomplish different missions through various pre-defined protocols. For example, the RFC1459 protocol enables IRC servers to communicate with client computers. Therefore, both hackers and network administrators must achieve the purpose of understanding the network operation mechanism through learning protocols.

Each protocol has been modified and used for many years. The newly generated protocols are mostly established based on the basic protocol. Therefore, the basic protocol has a relatively high-security mechanism. It is difficult for hackers to discover security problems in the protocol. However, for some new types of protocols, because of a short time and poor consideration, hackers for security reasons may also exploit them.

For community talk of network protocols, people think that the basic protocol used today has security risks at the beginning of the design. Therefore, no matter what changes are made to the network, if the network does not go under core changes, it is fundamentally impossible to impede any emergence of cyber hackers. However, this kind of hacking function is out of the confines of this book, and it is not covered here.

Second, the server and the client:

The most basic form of network service is several computers as clients, using a computer as a server where the individual client can send out requests to the server, then the server responds and completes the requested action, and finally, the server will return the execution result to the client computer. There are many such agreements. For example, the email server, web server, chat room server, etc. that we usually contact is all of this type. There is another kind of connection method where it does not need the server support, but directly connects two client computers; this makes the computers act as a server and client. Peer-to-peer completion of the connection and information exchange work. For example, the DCC transmission protocol falls into this category.

It can be seen from this that the client and the servers are the requesting application computer and the answering computer specified in various protocols, respectively. As a general Internet user, they all operate their own computers (clients) and send regular requests to the webserver to complete actions such as browsing the web, sending and receiving emails, and for hackers through their own computers (The client) attacks other computers (which may be clients or servers) to invade, destroy, and steal information.

Third, the system and system environment:

The operating system must be installed to operate the computer. The popular operating system is mainly UNIX, Linux, Mac, BSD, Windows2000, Windows95/98/Me, Windows NT, etc.. These operating systems run independently and have their own file management, memory management, process management, and other mechanisms. On the network, these different operating systems can be operated as servers or as clients, and they can exchange information through the protocol jobs.

Different operating systems and various applications constitute the system environment. For example, the Linux system can be used to configure the computer as a web server with Apache software. Other computers using the client can have the browser to get the website server for the viewer to read. The text information as Windows 2000 with Ftp software can be set up as a file server, through remote FTP login and can get various file resources on the system.

Fourth, IP address and port:

We go online and browse the web at the same time, send and receive an e-mail, voice chat, and many network services projects can be completed using various protocols, but the network is bigger than our computer. What do I do to find the computer I needed for my service? How to do so much work on one computer at the same time? Here we will introduce the IP address.

Computers connected with the Internet will have a unique IP address. An IP address is similar to a home address. Through various physical devices like network routers (without the need for newbies to understand). The computers in the network can easily do information exchange without any issues because their IP address is different; it is easier to find the target computer. Hackers, however, can make their computer's IP address change through specific methods, so any target server receives a request from the hacker. This is called Pseudo IP address. Servers will respond to the message sent from the pseudo IP address, thus causing network confusion. Hackers, of course, can find any surfers or servers based on IP addresses and attack them (think of real-time burglary) quickly.

Next, I will talk about the second question we talked about above: Why do I use multiple network services at the same time on one computer? It seems that New York City has eight gates. Distinct protocols will show in unique network services, and different network services will open via unique ports, much like City gates that help the client computer to complete its information transmission. In addition, if a web server has multiple network services open at the same time, it has to open a few different ports (city gates) to accommodate multiple and distinct client requests.

The back door that is often heard of on the Internet means that the hacker has opened up a network service on the server through specialized functions. The service hackers use to specifically complete their goals, and this will open with a new port. With this kind of service, regular internet users and administrators easily discover ports. These hidden ports are called a back door.

Each computer can open 65,535 ports. We can assume to develop at least 65,535 unique network services, but in fact, this number is very large. The network often uses dozens of service agreements, such as browsing web clients. Both port and server use port 80. For IRC chat, port 6667 is used on the server, and port 1026 is used on the client.

5) Vulnerabilities:

Vulnerabilities are situations that are not considered in the program. For example, the simplest "weak password" vulnerability means that the system administrator forgot to block accounts in some network applications. The Perl program vulnerability maybe because of the design of the programmers. When the program considers the imperfect situation, the code segment that causes the program to be executed is overwhelmed. The overflow vulnerability belongs to the original design of the system or program, without pre-reserving sufficient resources, and in the future, the program is used. The resulting resources are insufficient; the special IP packet bomb is actually an error when the program analyzes some special data, etc...

Overall, the loophole is a human negligence in the design of the program, which is almost improbable to avoid in any program, the hacker uses all kinds of loopholes to attack the network. The word "network security" at the beginning of this chapter is actually the meaning of "vulnerability." Hackers use vulnerabilities to complete various intrusion to get the ultimate result. In fact, hackers are really defined as "the person looking for vulnerabilities." They are not cyber-attackers for fun but are obsessed with getting in through other people's programs and looking for vulnerabilities every day. It is, to a certain extent, the hacker is the "good people." They are committed to this line in pursuit of perfection and establishment of a secure Internet, but only because some hackers or simply hackers often exploit aggressive vulnerabilities. In recent years, people have become scared of hackers.

6. Encryption and Decryption:

As an explanation of "Agreement," I cited "because of the problem of the grassroots of network design..." simply saying that this problem is to allow all users of Internet participating in information exchange, creating certain businesses, sharing personal privacy on the Internet will be exposed to participate in information sharing, and thus for certain businesses, the transmission of personal privacy on the Internet will be exposed to the public. Credit Cards, personal emails, etc. has the potential to be accessed by others through monitoring or interception. How can we make this information safe? The reader may have "World War II" thought of as spy war as the participating countries used the telegram to encrypt codes. Only the receiver who knows the password can decode the message. This ancient encryption method that still has its vitality in the modern network. The information processed by encryption is going through the network. No matter who gets the document, so long as they do not have a password, it is still in vain.

The longest use on the network is to set a personal password, use DES encryption lock, these two encryption methods can complete the user login system, website, email mailbox, and protection information package, and the work that hackers want to do is through loopholes. The brute force guessing, the reverse application of the encryption algorithm and other methods to obtain the plaintext of the encrypted file, some people use the "magic height one foot, and the road high one" is used here, it is indeed appropriate! Encryption methods on the network and systems that require password verification are emerging, and hackers are constantly looking for ways to break these systems.

It can be said that "vulnerabilities" and "decryption" are two completely different hacking fields. The preference of diverse learners for them will directly affect the types of hackers that they will become in the future, so the choices they make between them should be based on personal preferences, and this book will focus on learning about the "vulnerabilities."

Seventh, Trojan horse:

Trojan horse is an application designed and programmed by the programmer's intentional design. However, the operation of the Trojan horse, whether or not the user understands it, is not endorsed. According to some people's knowledge, viruses are a special case of Trojan horses: they can be spread to another program. They are also converted into Trojan horses. According to another person's understanding, viruses that are not intentionally causing any damage are not Trojan horses. Regardless of how people define it, in the end, many people only use "Trojan horses" to describe malicious programs that cannot be copied in order to distinguish Trojan horses from viruses.

Commonly Used Hacker Software Classifications

1. Prevention:

This is from a class of software involved in security perspectives, like firewalls, virus checking software, system process monitors, port management programs, etc., all of these belong to such software. This type of software maximizes and raises security and personal privacy for computer users and will not be compromised by hackers. Network servers give great importance to the needs of such software. Log analysis software, system intrusion software, etc. helps administrators in maintaining servers and track hackers who invade the system.

Second, information collection:

Information collection software types include port, vulnerability, and weak password scanning, and other scanning software, as well as monitoring, interception of information packets, and any spyware application, most of which belong to the software is also true and evil. That is to say, regardless of decent hackers, evil hackers, system managers, and ordinary computer users, user-like software can accomplish different purposes. In most cases, hacker-like software is more frequent because they rely on such software to scan the server in all directions, get more information about the server, and get a better understanding of the server. In order to carry out hacking.

3 Trojans and worms:

This software is different, but they work very much the same way, they are both virus-hidden and destructive, and such that this application is workable by the people with control or setup prior via well-designed procedures, but they do need a certain amount of work. Of course, this application is programmable for the use by system administrators as a remote management tool for servers.

4. Floods

The so-called "flood," that is, information garbage bombs, can cause the target server to overload and crash through a large number of garbage requests. In recent years, DOS distributed attacks have become popular on the network. Flood software may be used as a mail or chat bomb. These "fool" software has been streamlined and programmed by network security enthusiasts. Also, the software is often used in the hands of "pseudo-hackers" accused at the beginning of this book.

V. Password cracking:

The most practical way to ensure network security is to count on the cryptosystem of various encryption algorithms. Hackers have the ability to easily get ciphertext of the password file, but if there is an encryption algorithm, they still cannot obtain the real password. Therefore, the use of a password cracking application is imperative; using a computer with high-speed computing capabilities, software like these use dictionary password or an exhaustive way to restore the encrypted essay.

6. Deception:

When you need to get the plaintext password mentioned above, hackers need to perform encryption algorithm restoration on the ciphertext, but if it is a complicated password, it is not so simple to crack. However, is it more convenient to let the person who knows the password directly tell the prototype of the hacker password? Deception software is designed to accomplish this.

7. Camouflage:

The ISP and the server will record all kinds of processes and actions on the network. If the hacker's action is not performed after a good camouflage, it is easily tracked by any security technology, leading straight back to the hacker. So disguising own IP address and any identifying information is essential for hacker's compulsory course, but to use any camouflage technology requires deep expertise of the network. This kind of software is used when there is no solid foundation at the beginning.

The fourth important section you need to master is learning the basic environment of hackers.

First, they find the right operating system:

We usually hear hackers love Linux because Linux provides a far more flexible operation option with more powerful functions compared to Windows. Examples of these functions are the forgery of IP addresses, it is easy to write special IP header information using the Linux system, but it is almost impossible under Windows system. However, Linux also has its shortcomings. The commands in this system are complex and complicated, which makes it not convenient for new users. Individual learners will not be open to give up "comfortable" Windows, give up wonderful computer games and convenient operation, and go all out to hacker learning. In addition, new hackers get used to the Windows system as most of the knowledge of the network is to be learned there. Relative to the Linux system, the hacking software under the Windows platform is not infrequent. In addition, by installing the package, the Windows system can also be debugged. The amount of procedures, so the beginner hacker does not have to start with Linux.

This book uses the platform Kali Linux because, for individual users, NT or 2000 is a little more demanding - system configuration requirements are too high. However, the use of 95 or 98 lacks some of the necessary functions - NET, TELNET commands are not perfect. However, most of the contents of this book will evaluate vulnerabilities, starting from a remote server, so it really is not needed to learn Kali Linux operating system.

Second, the commonly used software:

If you are using a Kali Linux, then good news for you – you do not have to install extra software, because the hacking knowledge we will meet depends on the commands and built-in software provided by the system and can be done easily. Aside from the basic operating system, learners need to install a variety of scanners and get better Trojan software, monitoring software, and so on. When needed, readers may choose to install software above and learn how to use them, but I want to tell you that for all kinds of bombs, as well as a variety of hacking software on the network, after learning this book, you can if you make your own and develop it yourself, there will be no need of using software written by others when you have one developed by yourself.

For the scanner and monitoring software, I give the following suggestions, and the software will be described in detail later in the book:

All three of these software's are free and powerful. Like Nmap and Metasploit is a domestic software, it integrates a different scanning option that supports both console and graphical interface operations, as well as detailed vulnerability instructions. For beginners learning to hack these tools, are more than enough.

Third, additional tools:

If you are able to install the tools above, it would be of huge help to learn to jack; of course, the following software is mainly to acquire additional content and for the "second part" learning to pave the way, so it doesn't hinder the study of this book.

1. Background server:

A background service program with some application on the network can be programmed to make the computer like a small server to learn corresponding network applications and makes it easy to understand mechanical work internally, in turn, immensely improve its own server's perceptual knowledge, while also being able to monitor the data on its own server when the server is activated. If another hacker was to attack, you can clearly document the other party's attack process, which a beginner can learn more hacking methods. For this book, we mainly introduce scripting language vulnerabilities such as Perl and ASP, so we can install an IIS or HTTPD. Then set up Active Perl to make your own server to have the ability to compile CGI and pl scripts. There is also a benefit to using your own server. You save a lot of online time by putting all the processes of learning and finding vulnerabilities on your own computer, saving you money and poses no danger to any network.

2 C language compilation platform

In the future, when learning to hack, you will encounter many "problems of your own." Others may not notice these problems on the network, so you cannot find the corresponding program. At this time, it is a matter of developing the devices by yourself, so setting up Borland C++ will make it easier. Through this compiler, learners can learn both the C language and some of the small programs listed later in this book to create a Tool library.

Fourth, the classification of network security software

Now let us look at the kinds of network security applications because, as a learning hacker, knowledge is two interrelated processes: learning how to hack while preventing a hack is vital.

1. Firewall:

The most common security application set up on any network. The firewall has both hardware and software. Most readers may see software firewalls. Its functions are mainly to filter spam (this is to make sure that your system will not be bomb attacked), to prevent any intrusions, whether by employing worms or hacking, to elevate the system's privacy to protect sensitive data, to monitor system resources in real-time, to prevent system crashes, and to maintain databases regularly. Backing up the main information... The firewall can patch vulnerabilities any system may have, leaving the hacker no chance even to try. In addition, for enterprises with LANs, firewalls can limit the opening of system ports and prohibit specific network services (to prevent Trojans).

2. Detection software:

The internet has a device for clearing a hacker program. The application, however, is combined with the Firewall and anti-virus software installed. If Trojans and worms are detected in the system and cleared, the software, in order to ensure there is no system infringement, it will automatically protect the hard disk data, automatically maintains the registry file, detect the content of the code, and monitor the open status of the system port. If the user wants, they can set up a script in the software to shield a specified port (this function is the same as the firewall).

3. Backup tools:

These are applications meant to make a copy of the data in a server, which helps to update the data at the time of development, so even if and when a hacker destroys the database on the server, the software can completely repair the received intrusion data in a short time. In addition, for individual users, this kind of software can do a full image backup of the hard drive that, in the event of a system crash, users can restore the system to its original state at a certain point. An example of this is a software called Ghost.

4. Log records, analysis tools:

For a server, the log file is quintessential, as this is the tool that helps the administrator to check what requests the server has been receiving and where it was sent. This allows administrators to know when they have been hacked definitively, and with the help of the log analysis software, they can easily set up trackers for any intrusion, find where the hacker entered the system, and then find the hacker's location this way. For this very reason, hackers have to learn how to do IP address masquerading, server hopping, and clearing log files after hacking a server.

Installing a Virtual Machine

People must be prepared for everything. Hackers are no exception. Before hackers invade other computers on the Internet, they need to do a series of preparations, including installing virtual machines on computers, preparing commonly used tools, and mastering common ones.

Whether it is an attack or training, hackers will not try to use a physical computer, but build a virtual environment in a physical computer, that is, install a virtual machine. In a virtual machine, hackers can intuitively perform various attack tests and complete most of the intrusion learning, including making viruses, Trojans, and implementing remote control.

A virtual machine is a computer system that is simulated by software and mimics a system with complete hardware functionality and functions as an independent environment. The work that can be done on the physical machine can be implemented in the virtual machine. Because of this, more and more people are using virtual machines.

When you create a new virtual machine on a computer, you need to use part of the hard disk and memory capacity of the physical machine as the hard disk and memory capacity of the virtual machine. Each virtual machine has its own CMOS, hard drive, and operating system. Users can partition and format the virtual machine, install operating systems and application software, just like a physical machine.

The Java Virtual Machine is an imaginary machine that is typically implemented by software simulation on a real computer. The Java virtual machine has its own imagined hardware, such as processors, stacks, registers, etc., and has a corresponding instruction system. The Java virtual machine is mainly used to run programs edited by Java. Because the Java language has cross-platform features, the Java virtual machine can also directly run programs edited in Java language in multiple platforms without modification. The relationship between the Java virtual machine and Java is similar to the relationship between Flash Player and Flash.

There may be users who think that the virtual machine is just an analog computer, and at most, it can perform the same operations as a physical machine, so it does not have much practical significance. In fact, the biggest advantage of a virtual machine is virtualization. Even if the system in the virtual machine crashes or fails to run, it will not affect the operation of the physical machine. In addition, it can be used to test the latest version of the application or operating system. Even if the installation of the application with the virus Trojan is no problem because the virtual machine and the physical machine are completely isolated, the virtual machine will not leak in the physical machine data.

VMware is a well-known and powerful virtual machine software that allows users to run two or more windows and Linux systems simultaneously on the same physical machine. Compared with the "multi-boot" system, VMware adopts a completely different concept. Multiple operating systems of a physical machine can only run one of the systems at the same time. The switching system needs to restart the computer, but VMware is different. It is the same. Multiple operating systems can be run at any time, thus avoiding the hassle of rebooting the system.

The VMware installer can be downloaded from some common resource offering sites such as filehippo.com. After downloading the VMware installer, you can extract and install it. After the installation is successful, the corresponding shortcut icon will be displayed on the desktop.

The following describes the steps to create a new virtual machine in VMware.

STEP01:

Start VMware Workstation by using the GUI interface.

STEP02:

Select a new virtual machine

STEP03:

 Select the configuration type

STEP04:

 Select to install the operating system later

STEP05:

Select the guest operating system

STEP06:

Set the virtual machine name and installation location

STEP07:

 Specify virtual machine disk capacity

STEP08:

Click the "Finish" button

Installation of Kali Linux

Nowadays, the installation process of Linux has been very "fast," and the installation of the entire system can be completed with a few mouse clicks. The installation of the Kali Linux operating system is also very simple. This section describes the detailed process of installing Kali Linux to the hard drive, USB drive. We will explain how to upgrade tools in the next section.

Installing to a hard drive is one of the most basic operations. The implementation of this work allows users to run Kali Linux without using a DVD. Before you install this new operating system, you need to do some preparatory work. For example, where do you get Linux? What are the requirements for computer configuration? ... These requirements will be listed one by one below.

The minimum disk space for Kali Linux installation is 8GB. For ease of use, it is recommended to save at least 25GB to save additional programs and files.

The memory is preferably 512MB or more.

The official website provides 32-bit and 64-bit ISO files. This book uses 32-bit as an example to explain the installation and use. After downloading the ISO file, burn the image file to a DVD. Then you can start to install Kali Linux to your hard drive.

(1) Insert the installation CD into the CD-ROM of the user's computer, restart the system, and you will see the interface

(2) This interface is the guiding interface of Kali, and the installation mode is selected on this interface. Selecting the Graphical Install here will display an interface.

3) Select the default language of the installation system in this interface as English, and then click the Continue button then the next interface will be shown.

(4) In the interface selection area is "Your country," and then click the "Continue" button, the next interface will be displayed.

(5) Select the keyboard mode as "English" in this interface, and then click "Continue" button, the next interface will be displayed.

(6) This interface is used to set the hostname of the system. Here, the default hostname Kali is used (users can also enter the name of their own system). Then click the "Continue" button, the next interface will be displayed.

(7) This interface is used to set the domain name used by the computer. The domain name entered in this example is

kali.example.com. If the current computer is not connected to the network, you can fill in the domain name and click the "Continue" button. The next interface will be displayed.

(8) Set the root user password on this interface, and then click the "Continue" button, the next interface will be displayed.

(9) This interface allows the user to select a partition. Select "Use the entire disk" here, and then click the "Continue" button, the next interface will be displayed.

(10) This interface is used to select the disk to be partitioned. There is only one disk in the system, so the default disk is fine here. Then click the "Continue" button, the next interface will be displayed.

(11) The interface requires a partitioning scheme, and three schemes are provided by

default. Select "Place all files in the same partition (recommended for beginners)" and click the "Continue" button, the next interface shown will be displayed.

(12) Select "Partition setting ends and writes the changes to disk" in the interface, and then click "Continue" button, the next interface will be displayed. If you want to modify the partition, you can select "Undo the modification of the partition settings" in this interface to re-partition.

(13) Select the "Yes" checkbox on this interface, and then click the "Continue" button, the next interface will be displayed.

(14) Start installing the system now. Some information needs to be set during the installation process, such as setting up network mirroring. If the computer on which the Kali Linux system is installed is not

connected to the network, select the "No" checkbox on this screen and click the "Continue" button. Select the "Yes" checkbox here, and the next interface will be displayed.

(15) Set the HTTP proxy information on this interface. If you do not need to connect to the external network through the HTTP proxy, just click the "Continue" button, the next interface will be displayed.

(16) After the scanning mirror site is completed, you can go to the next option

(17) In the country where the image is selected, select "Your country" and click "Continue" button, the next interface will be displayed.

(18) The interface provides 7 mirror sites by default, and one of them is selected as the mirror site of the system. Select

mirrors.163.com here, then click the "Continue" button, the next interface will be displayed.

(19) Select the "Yes" checkbox on this interface, and then click the "Continue" button, the next interface will be displayed.

(20) The installation will continue at this time. After the installation process is finished, Kali Linux login screen will appear.

Installing kali Linux using a USB drive

The Kali Linux USB drive provides the ability to permanently save system settings, permanently update and install packages on USB devices, and allows users to run their own personalized Kali Linux. Create a bootable Live USB drive for the Linux distribution on the Win32 Disk Imager, which includes continuous storage for Kali Linux.

This section describes the steps to install Kali Linux to a USB drive.

Installing an operating system onto a USB drive is a bit different from installing to a hard drive. Therefore, you need to do some preparation before installing it. For example, where do you get Linux? USB drive format? What is the size of the USB drive? These requirements will be listed one by one below.

After the previous preparations are completed, you can install the system. The steps to install Kali Linux onto a USB drive are as follows.

(1) Insert a formatted and writable USB drive into the Windows system. After inserting, the display next interface is shown.

2) Start Win32 Disk Imager, the startup interface is shown. In the Image File location,

click the icon to select the location of the Kali Linux DVD ISO image and select the USB device where Kali Linux will be installed. The device in this example is K. After selecting the ISO image file and USB device, click the Write button to write the ISO file to the USB drive.

(3) Use the UNetbootin tool to make the device K a USB boot disk. Launch the UNetbootin tool, and the next interface will be displayed.

(4) Select the "Disc Image" checkbox in this interface, then select the location of the ISO file and set the Space used to preserve files across reboots to 4096MB.

(5) Select the USB drive, the USB drive in this example is K, and then click the "OK" button; it will start to create a bootable USB drive.

(6) After the creation is completed, the next interface will be displayed.

(7) At this point, the USB drive is created successfully. In the interface, click the "Restart Now" button, enter the BIOS boot menu and select USB boot, you can install the Kali Linux operating system.

When users use it for a while, they may be dissatisfied with working in a system that does not change at all but is eager to upgrade their Linux as they would on a Windows system. In addition, Linux itself is an open system, new software appears every day, and Linux distributions and kernels are constantly updated. Under such circumstances, it is very important to learn to upgrade Linux. This section will introduce Kali updates and upgrades.

Updating and Upgrading Kali Linux

The specific steps for updating and upgrading Kali are as follows.

(1) Select "Application" | "System Tools" | "Software Update" command in the graphical interface, and the next interface will be displayed.

(2) The interface prompts to confirm whether the application should be run as a privileged user. If you continue, click the "Confirm Continue" button, the next interface will be displayed.

(3) The interface shows that a total of packages need to be updated. Click "Install Update" button to display the interface.

(4) This interface shows the packages that the update package depends on. Click the "Continue" button to display the interface.

(5) From this interface, you can see the progress of the software update. In this interface, you can see a different status of each package. Among them, the package appears behind the icon, indicating that the package is downloading; if displayed as icons indicate the package has been downloaded; if there is at the same time and icon, then, after you install this package, you need to reboot the system; these packages are installed once successful, it will appear as an icon. At this point, click the "Exit" button and restart the system. During the update process, downloaded packages will automatically jump to the first column. At this point, scrolling the mouse is useless.

(6) After restarting the system, log in to the system and execute the lsb_release -a command to view all version information of the current operating system.

7) From the output information, you can see that the current system version is 2.2.1. The above commands apply to all Linux distributions, including RedHat, SuSE, and Debian. If you only want to view the version number, you can view the /etc/issue file. Execute the command as follows:

root@kali:~# cat /etc/issueKali GNU/Linux 2.2.1\n \l

A Hacking Roadmap

If a hacker wants to attack a target computer, it cannot be done by DOS commands. It also needs some powerful intrusion tools, such as port scanning tools, network sniffing tools, Trojan making tools, and remote-control tools. This section will briefly introduce the intrusion tools commonly used by hackers.

a) Port scanning

The port scanning tool has the function of scanning the port. The so-called port scanning means that the hacker can scan the information of the target computer by sending a set of port scanning information. These ports are intrusion channels for the hacker. Once the hacker understands these ports, the hacker can invade the target computer.

In addition to the ability to scan the open ports of a computer, the port scan tool also has the ability to automatically detect remote or target computer security vulnerabilities. Using the port scan tool, users can discover the distribution of various TCP ports on the target computer without leaving traces. In addition, the services provided to allow users to indirectly or directly understand the security issues of the target computer. The port scanning tools commonly used by hackers are SuperScan and X-Scan.

b) Sniffing tool

A sniffing tool is a tool that can sniff packets on a LAN. The so-called sniffing is to eavesdrop on all the packets flowing through the LAN. By eavesdropping and analyzing these packets, you can peek at the private information of others on the LAN. The sniffing tool can only be used in the local area network, and it is impossible to directly sniff the target computer on the Internet. The data sniffing tools commonly used by hackers are Sniffer Pro and Eiffel Web Detective.

3) Trojan making tool

As the name suggests, Trojan making tools are tools for making Trojans. Since Trojans have the function of stealing personal privacy information of the target computer, many junior hackers like to use Trojans to make Trojans directly. The Trojan creation tool works basically the same way. First, the tool is used to configure the Trojan server program. Once the target computer runs the Trojan server program, the hacker can use the Trojan tool to completely control the target computer of the Trojan.

The operation of the Trojan making tool is very simple, and the working principle is basically the same, so many junior hackers favor it. Trojan horse making tools commonly used by hackers are "glacial" Trojans and bundled Trojans.

4) Remote control tools

Remote control tools are tools with remote control functions that can remotely control the target computer, although the control methods are different (some remote-control tools are remotely controlled by implanting a server program, and some remote-control tools are used to directly control the LAN, and all computers in the middle), but once the hacker uses the remote-control tool to control the target computer, the hacker acts as if he/she were sitting in front of the target computer. The remote-control tools commonly used by hackers are network law enforcement officers and remote control.

Hacking Target Computers

On the Internet, to prevent hackers from invading their own computers, it is necessary to understand the common methods of hacking target computers. The intrusion methods commonly used by hackers include data-driven attacks, illegal use of system files, forged information attacks, and remote manipulation. The following describes these intrusion methods.

1) A data-driven attack

A data-driven attack is an attack initiated by a hacker who sends or copies a seemingly harmless unique program to a target computer. This attack allows hackers to modify files related to network security on the target computer, making it easier for hackers to invade the target computer the next time. Data-driven attacks mainly include buffer overflow attacks, format string attacks, input verification attacks, synchronous vulnerability attacks, and trust vulnerability attacks.

2) Forgery information attack

Forgery information attack means that the hacker constructs a fake path between the source computer and the target computer by sending the forged routing information so that the data packets flowing to the target computer are all passed through the computer operated by the hacker, thereby obtaining the bank account in the data packet—personal sensitive information, such as passwords.

3) Information protocol

In a local area network, the source path option of the IP address allows the IP packet to choose a path to the target computer itself. When a hacker attempts to connect to an unreachable computer A behind a firewall, he only needs to set the IP address source path option in the sent request message so that one of the destinations addresses of the packet points to the firewall, but the final address points to Computer A. The message is allowed to pass when it reaches the firewall because it points to the firewall instead of computer A. The IP layer of the firewall processes the source path of the packet and sends it to the internal network. The message arrives at the unreachable computer A, thus achieving a vulnerability attack against the information protocol.

4) Remote operation

Remote operation means that the hacker launches an executable program on the target computer. The program will display a fake login interface. When the user enters the login information such as account and password in the interface, the program will save the account and password then transfer it to the hacker's computer. At the same time, the program closes the login interface and prompts the "system failure" message, asking the user to log in again. This type of attack is similar to a phishing website that is often encountered on the Internet.

5) LAN security

In the local area network, people are one of the most important factors of LAN security. When the system administrator makes a mistake in the configuration of the WWW server system and the user's permission to expand the user's authority, these mistakes can provide opportunities for the hacker. Hackers use these mistakes, plus the command of a finger, netstat, etc., to achieve intrusion attacks.

Resending an attack means that the hacker collects specific IP data packets and tampers with the data, and then resends the IP data packets one by one to spoof the target computer receiving the data to implement the attack.

In the LAN, the redirect message can change the router's routing list. Based on these messages, the router can suggest that the computer take another better path to propagate the data. The ICMP packet attack means that the hacker can effectively use the redirect message to redirect the connection to an unreliable computer or path or to forward all the packets through an unreliable computer.

6) Vulnerability attack

A vulnerability attack for source path selection means that the hacker transmits a source path message with an internal computer address to the local area network by operating a computer located outside the local area network. Since the router will trust this message, it will send an answer message to the computer located outside the LAN, as this is the source path option requirement for IP. The defense against this type of attack is to properly configure the router to let the router discard packets that are sent from outside the LAN but claim to be from internal computers.

7) Ethernet broadcast attack

The Ethernet broadcast attack mode refers to setting the computer network card interface to promiscuous, to intercept all the data packets in the local area network, analyze the account and password saved in the data packet, and steal information.

UNIX

On the Internet, servers or supercomputers on many websites use the UNIX operating system. The hacker will try to log in to one of the computers with UNIX, get the system privilege through the vulnerability of the operating system, and then use this as a base to access and invade the rest of the computer. This is called Island-hopping.

A hacker often jumps a few times before attacking the final destination computer. For example, a hacker in the United States may log in to a computer in Asia before entering the FBI network, then log in to a computer in Canada, then jump to Europe, and finally from France. The computer launches an attack on the FBI network. In this way, even if the attacked computer finds out where the hacker launched the attack, it is difficult for the administrator to find the hacker. What's more, once a hacker gains the system privileges of a computer, he can delete the system log when exiting and cut the"vine."

In almost all protocol families implemented by UNIX, a well-known vulnerability makes it possible to steal TCP connections. When a TCP connection is being established, the server acknowledges the user request with a response message containing the initial sequence number. This serial number has no special requirements, as long as it is unique. After the client receives the answer, it will confirm it once, and the connection will be established. The TCP protocol specification requires a serial number of 250,000 replacements per second, but the actual replacement frequency of most UNIX systems is much smaller than this number, and the number of next replacements is often predictable, and hackers have this predictable server initial. The ability of the serial number allows the intrusion attack to be completed. The only way to prevent this attack is to have the starting sequence number more random. The safest solution is to use the encryption algorithm to help generate the initial sequence number. The resulting extra CPU load is now the hardware speed. It can be ignored.

On UNIX systems, too many files can only be created by super users, and rarely by a certain type of user. This makes it necessary for system administrators to operate under root privileges. This is not very safe. Since the primary target of hacking is the root, the most frequently attacked target is the super user's password. Strictly speaking, the user password under UNIX is not encrypted. It is just a key for encrypting a common string as a DES algorithm. There are now a number of software tools for decryption that use the high speed of the CPU to search for passwords. Once the attack is successful, the hacker becomes an administrator on the UNIX system. Therefore, the user rights in the system should be divided, such as setting the mail system administrator management, and then the mail system mail administrator can manage the mail system well without superuser privileges, which makes the system much safer.

Chapter 7: Bash and **Python** scripting

This chapter will give an excellent introduction to bash, a command-line interface language, and python, a popular programming language. By learning bash and python as a hacker, you can increase your skills exponentially. You may feel overwhelmed looking at all of the code that you might have never seen before. Try to practice code by doing little projects and automate tasks with python you will become an experienced hacker in a very short time. Different sections in the chapter will help you understand things easily and will let you learn to script effectively. Let us start!

What is a shell?

In Windows systems, we have a GUI that helps us to run programs. Of course, Linux based systems also consist of a GUI. However, apart from GUI Linux based systems consist of a powerful interface called shell a command-line interface. Shell helps the users to run a program or software using a command prompt. The shell can be executed directly using commands or by a file called shell scripts that can be easily created by a text editor or an IDE.

What is UNIX?

UNIX is an operating system that Linux and many operating systems are based upon. Learning about UNIX history can help us learn about the importance of shell in the programming world.

What is the bash?

There are many types of shell types. Among them, Bash is one of the most familiar UNIX shells that is simple and can be used to automate many tasks in the system.

What is the terminal?

The terminal is just like a browser for websites. It is developed for the client and for his comfort. People use terminals to type commands and start shell processing. Every operating system has a command terminal. For example, even windows have a terminal called MS-DOS, which has a lot of difference from the Linux terminals.

Looking at a terminal

Linux terminals are very easy to understand. When starting the terminal is a user in Kali

Linux we can see "userofthesystem@machinename" format followed by a $.

Before going to learn about the bash in detail, we will just go through a few examples that will explain how bash works.

1) Echo

You might have already heard it many times. This just displays whatever written inside it as the output.

$echo ' That's a good bash terminal'

output:

That is a good bash terminal

2) Date

This bash command displays the present date and time as output. This is used very commonly while doing scripting.

$date

Output:

Sun Dec 20 23:32:12 PST 2019

3) Calendar

$cal

This will just display the calendar of the month that you are in.

We will give a simple example by creating a shell file instead of directly executing it in the command. This process is explained below in detail.

Step 1:

In the first step open a text editor and create a file named example.sh and start writing the following lines.

!/bin/bash

echo "This happens in every shell" // this prints as output

We will explain the shell program line by line now. The first line will just order the file to open in a bash shell. Also, the second line, as we already discussed, will print the text in between quotes as output. The next (beside the command) is a text that can be written to help the reader or programmer.

Step 2:

Save the file. In the next step, to execute the following shell script file, we need to modify the permission. This can be done using the following command.

chmod +x example.sh // This will create executable permission to the script

Step 3:

In the last step, we need to execute the file in the command to get an output. We can use the following commands to get an output.

$ bash example.sh

$ ./example.sh

Output:

This happens in every shell

We can use the command Clear to clear everything on the shell.

$ clear

In the next section, we will explain some basic bash commands that are available in Kali Linux. These are very important to learn for a better understanding of the Linux system functions.

1) pwd

This command will print the present working directory that the user is in. You can even look at a tree directory with additional options.

$ pwd

Output:

/user/rod

2) ls

This command will show the present directory contents that are all of its files and folders. We normally use it to look at the present directory files.

$ ls

Output:

shell.py shell.img   shell.png

3) cd

This command is used to change the directory. Get ready with the path that you want to navigate to and add cd before it. You can check whether the directory is changed or not by using ls command.

$ cd /home/desktop/

4) mkdir

This command can be used to make a new directory or new folder.

```
$ mkdir   new/ruby
```

5) mv

This command helps us to move files or folders from one directory to another. It follows the following pattern.

```
$ mv  sourcepath    destinationpath
```

For example:

```
$ mv    plus.py plus/code
```

6) touch

This is a special command that can be used to create a new file of any type in a director. For example, let us have an empty text file called example.py in the present directory.

$ touch example.txt

7) rm

This command helps us to remove a file from the disk. This will completely delete the file. So use with caution.

$ rm example.txt

8) rmdir

This command removes the directory from the disk.

rmdir python/python files

9) cat

This command can be used to read the text file and display everything present on in the output screen.

$ cat rowdy.txt

Output:

Apple

Bat

Car

Dog

.

.

.

This will display everything that is present in the rowdy.txt text file.

However, when trying to display a text file with a large number of data, it can become quite clumsy. So, to get rid of this, we use the command below.

10) less

This command displays a huge chunk of data in one page per time. This will help us by using certain gestures. Spacebar can take us to the next page, whereas b will take us to the previous page.

$ less rowdy.txt

In the next section, we will describe important concepts called pipelines and filters that can be used for organization and other purposes.

1) Pipelines

This explicitly means to give the output of the first command as an input to the second command.

command1 | command2

Where | is called as a pipe operator.

2) Filters

Filters are extra techniques used to separate or organize data. Linux and bash consist of many filter commands to find and organize data.

a) Grep

This command helps to find a word in a text file easily. The command is as follows to understand the functionality of grep command.

$ grep raj cricket.txt

This will display everything that matches raj in the text file.

b) Sort

This command will help to sort the contents in the file alphabetically or numerically.

Example for pipelines & filters:

```
$cat animals.txt | sort  // $ command textfile
pipeline(|) filter
```

Output:

Cat

Dog

Monkey

Zebra

Now we have mastered all the basic bash properties and now will dive into more complex topics that will explain the bash in depth.

1) Variables

Variables are an important piece of memory that stores the data given as input or while executing. The data to be processed should contain variables that have different data types. In short, the variable is like an address box and can be modified or replaced with another variable if given correct instructions.

Bash also consists of variables and can be used to write command-line instructions that can-do various tasks.

Bash variables are quite different from other variables because they are of two types. Below we will explain them in detail.

1) System variables

These variables are pre-created by Linux and can be used while scripting to fill a particular value. These are represented by capital letters.

Example:

USERNAME ----> This variable defines the current logged in user name. You can call it to get the specific output.

2) User variables

These are variables, which are user-generated and can be used for complex tasks. Lower case letters represent these. In bash, variables can store any data irrespective of datatype we normally use in programming languages.

a) Define a variable

# Here is the syntax
$ room = bad

b) To use a variable

$room

c) Now you can use this command to print the following variable using the echo command.

echo $room

Output:

bad

d) We can also use this variable in a string for printing it. This is quite easy and very useful function.

$ echo " This is $room"

output:

This is bad

Conditionals in the Bash

This is normal if and else statement that is used in programming languages. We will explain this with an example below.

```
# This is a bash script to describe conditioning

if [15 -lt 25]
then
   echo "This is smaller"
fi
```

output :

This is smaller

In the above example, we just explained a statement, and in the next example, we will use a model with an else statement too.

This explains both if and else

```
if      [      35      -lt      25      ]
then
 echo   "This   is   the   biggest   number"
else
 echo   "This   is   the   lesser   number"
fi
```

Output:

This is the biggest number

Looping in a shell script:

Looping means to repeat the same thing with a definite interval. Bash has for loop and while loop. To understand this precisely, we will use the following example.

bash program for loop

```
for j in 6 7 8 9
do
echo "This is $j"
done
```

Output :

This is 6

This is 7

This is 8

This is 9

Functions in a shell script:

A function is a set of instructions that need to be followed in a definite way. Bash provides many inbuilt functions, and we can create user-made functions as shown below.

syntax to create a function

```
addition()
{
  echo ' Sum is $a + $b'
  return
}
```

Now we need to call that function

```
addition
```

Output:

If variables are 2 and 3, we will get the output as

Sum is 5

This ends our journey to the world of shell scripting. There are many more bash scripting patterns and commands that you need to master to be a proficient hacker.

Chapter 8: Basic kali Linux concepts

Analyzing and managing networks

Hackers always tend to do quite complex things that can be tracked easily by forensic investigation. All major companies try to deploy forensic specialists and security investigators to find the details of the attacker after an attack. This means that being a hacker is not easy, and one who aspires to be a hacker needs to know a lot about networking and its management like spoofing his IP or physical mac address. This section will help us to learn in detail about these techniques in Kali Linux for a better attack probability.

ifconfig

This is a basic network command that is used in all Linux distributions to check the connected networks with the computer. You can find both wired and wireless connections using this command.

To use this command, you need to have root privileges as it contacts with the kernel to get more information about the network devices that are connected.

The command is below:

root@kali : ifconfig

After clicking the above command, you will get an output that displays the network devices that are connected. If there is an Ethernet-based connection that is wired, they will be represented as eth followed by a subscript of a number that starts from 0 like eth0, eth1, eth2, and so on. You will also find MAC and Ip address of the particular network.

We can use this information while attacking to make things difficult to find.

ifconfig also displays wlan0 that is about a wireless adapter that is within your system. "Hwaddr" also called as MAC address is displayed and this can be widely used in aircrack-ng while you are trying to attack a WIFI access point. The next section will describe how to know about wireless connections in detail.

Didn't

In your Kali Linux terminal as a root user, click the following command to know more about wireless interfaces that are connected to the system.

root @ kali : iwconfig

This will display the wireless connection along with its encryption like WPA, WEP, and its physical address. For any hacker who is trying to capture packets for gaining sensitive information, a good overview of wireless adapter can make the process more interesting.

How to change the IP address?

A very basic idea that everyone knows is every network connection is distinguished by an address called an IP address. It is easy to track down the information if someone obtains your network address. In addition, for hackers who always try to attack hosts, this would be a problem. However, do not worry because there are few techniques and tricks, which you can use to spoof your address while doing a password attack or DOS attack. The below section will help you get more information about this process.